Principles of Redox Reactions

By Malika Ammam, PhD

Copyright© 2017 Malika Ammam. All rights reserved.

Discount Offers

5% OFF of the book price for purchases of 1-5 books.

8% OFF of the book price for purchases of more than 5 books.

To receive the discount money, send your request through https://www.malika-ammam.com/ with your order details and PayPal account. Make sure that your order details (amazon or other sites) passed the 30 days return policy.

Thank you,

Introduction

As a teacher of physical chemistry, I noticed that students, even in advanced classes, have difficulties in understanding the basics of redox chemistry. In this Section 3, I attempted to discuss some fundamental principles related to redox processes, by focusing on the species that might lose or gain electrons, determination of the oxidation numbers (or states) of atoms in compounds, and ways of balancing redox reactions. To further clarify the discussed concepts, numerous questions and problems with detailed answers are provided. Most of these questions are formulated by students like you. I believe that this Section 3 would greatly help students with levels varying from high school to advanced university classes.

Abstract

After reading Section 2, students should be able to identify the electronic structure of any element of the periodic table and be able to visualize the electrons of the outer shell, which could be involved in redox processes. This section 3 summarizes the basics of redox (or electrochemical) processes by focusing on the species that might lose or gain electrons, determination of the oxidation numbers (or states) of atoms in species involved in redox processes, and easy ways of balancing redox reactions.

1. Electrochemical (or redox) reactions

Electrochemical (or redox) processes can be defined as chemical reactions involving the exchange of electrons[1-2], which could be reduction or oxidation. A reduction refers to a process where a molecule, atom or ion gains one or more electrons and oxidation undergoes the opposite process where one or more electrons are lost during the reaction process. Remember that the electrons involved in these processes are the valence electrons located in the outer shell. On the other hand, though both oxidation and reduction half-reactions could be expressed independently, electrochemical processes always occur together where the electrons lost by one species are gained by another. This induces some kind of flow of electrons from one species to another, referred to as electron transfer.

The redox reaction of a substance A undergoing an oxidation with m electrons (me^-) with a substance B subjected to a reduction with n electrons (ne^-) could be summarized by Eqs. (1) and (2). A and B could be any element from the periodic table or a combination of atoms forming a substance or molecular compound.

Oxidation half-reaction: $(A \rightarrow A^{m+} + me^-) \times n$ (1)

Reduction half-reaction: $(B + ne^- \rightarrow B^{n-}) \times m$ (2)

Overall reaction: $nA + mB + mne^- \rightarrow nA^{m+} + mB^{n-} + nme^-$ (1+2)

Note that the charges of A and B changed during the oxidation and reduction processes. A becomes positive after losing electrons and B negative after gaining electrons. The oxidation and reduction reactions are called redox half-reactions, and their sum gives an overall reaction. The electrons generated during oxidation are consumed during reduction, always canceling the electrons in the overall reaction. That is why both the oxidation and reduction reactions are multiplied by the factors (n and m) to equalize the number of electrons on both sides of the overall reaction. Note that redox reactions might involve the transfer of one or several electrons, depending on the reacting substances, valence electrons, and experimental conditions.

2. Oxidation state (or number)

The oxidation state (or number) refers to the change in valence electrons of species, which could be gained or donated during a redox process[3-8]. If an atom is oxidized (or lost electrons), its oxidation number will increase. By contrast, if an atom is reduced (or gained electrons), its oxidation number will decrease. Note that the oxidation numbers are hypothetical and have no physical meaning but useful to figure out the number of electrons involved in redox processes, as well as facilitate the balancing of the involved equations. Oxidation number (or state) represents essentially the electric charge of an atom as if it is dissociated from the compound formula by taking into consideration the electronegativity of each part of the formula. Therefore, redox reactions are chemical reactions characterized by changes in oxidation states of at least two atoms.

3. Identification of oxidizing and reducing agent

Substances containing oxidized atom(s), which lost electrons, become oxidizing agents (also called, oxidants or oxidizers). Because of the lack of electrons in their valence subshells, oxidizing agents have the ability to oxidize other compounds by removing electrons from their outer shells and become reduced. By contrast, substances containing reduced atom(s) become reducing agents (also called, reductants or reducers). Since reducing agents have extra electrons in their valence subshells, they have the ability to reduce other substances by giving away electrons and become oxidized. This means that during redox processes, oxidizing agents become reducing agents and vice versa[3-8]. In other words, every reducing agent donating electron(s) becomes an oxidizing agent, and in turn, every oxidizing agent accepting electron(s) becomes a reducing agent. In Eq. (1), A^+ is the oxidant and A is the reductant, and both A^+ and A form a redox couple, denoted as A^+/A. This notation is very practical in indicating which species are oxidized or reduced, and allowing the correct identification of the oxidation and reduction half-reactions.

Elements with several oxidation states can either be oxidizing or reducing agents at intermediate states. For example, Fe^{2+} can play the role of an oxidizing agent in some reactions and reducing agent in others.

Fe^{2+} as oxidizing agent: $Fe \rightarrow Fe^{2+} + 2e^-$

Fe^{2+} as reducing agent: $Fe^{2+} \rightarrow Fe^{3+} + 1e^{-}$

4. Assigning an oxidation state (or number)

The electronic configuration of elements discussed in the previous section helps in determining the oxidation state of species involved in redox reactions. Assigning an oxidation number (or state) to an element in a substance (or compound) should always begin with elements having stable oxidation numbers, described by the rules summarized below[1-8].

- The first rule states that uncombined or free elements should have oxidation numbers of zero. For example, Na, Au or Fe all have oxidation numbers of zero because they are at uncombined or free states.

- The second rule suggests that monoatomic ions should have oxidation numbers equal to their formal charges. For instance, the oxidation number of Na^+ is +1, Mg^{2+} is +2, and S^{-2} is -2.

- The third rule states that oxygen (O) often has an oxidation number of -2, such as in NO and CO_2. Some exceptions to this rule exist, such as in hydrogen peroxide H_2O_2 (-1) and when combined with F in OF_2 (+2).

- The fourth rule indicates that hydrogen in nonionic compounds should have an oxidation number of +1, such as in CH_3, HCl, and H_2O. Some exceptions to this rule also exist when H is combined with metals, such as NaH or MgH_2 (-1).

- The fifth rule proposes that fluorine (F) should always have an oxidation number of -1, such as in NaF and KF.

- Rule number six states that elements of group 1 in the periodic table (Li, Na, K, Rb, Cs, and Fr) often have oxidation states of +1, those of group 2 (Be, Mg, Ca, Sr, Ba, and Ra) have oxidation states of +2, and elements of group III (Sc, Y, La, and Sc) have oxidation states of +3.

- Rule number seven proposes that oxidation states of nonmetals involving no oxygen or hydrogen in their structures depend on their electronegativities. Elements with higher electronegativities should have oxidation states equal to their commonly known negative ions. For example, F in SF6 has higher electronegativity value than S. Thus, the oxidation number of F is – 1 and that of S is +6. S in CS_2 has superior electronegativity than C. Hence, the oxidation number of S is –2 and that of C is +4. Note that this rule is no longer valid for elements with similar electronegativities, such as N_4S_4.

- The rule number eight suggests that the algebraic sum of all oxidation numbers should give the overall charge of the substance. For neutral molecules or compounds, the sum is zero.

For charged ionic compounds, the sum should give the overall charge (positive or negative). For instance, the overall charge of $(NH_4)(NO_3)$ is zero, thus the compound is neutral. In that case, the oxidation number could be determined separately in each component: $(NH_4)^+$ and $(NO_3)^-$. The oxidation number of H in $(NH_4)^+$ is +1 and that of N is -3, thus $-3 + 4(+1) = +1$. Similarly, the oxidation number of O in $(NO_3)^-$ is -2 and that of N is +5, thus $+5 + 3(-2) = -1$.

- The last rule proposes that the total oxidation number should always be preserved during chemical reactions. This allows distinguishing redox from chemical reactions, where oxidation raises the oxidation number and reduction decreases it.

5. Balancing redox reactions

Balancing redox reactions is more complex than chemical reactions because of the additional electron transfer process. Two main methods are often utilized for balancing redox reactions. The first is based on the oxidation number and the second on the half-reaction method[1-8].

5.1. Oxidation number method

The oxidation number (or state) could be utilized to balance redox reactions following these steps.

- The first step consists of identifying the elements subjected to changes in oxidation numbers during the redox process.
- Next, the total number of electrons lost by the reductants should be equal to the electrons gained by the oxidants.
- Afterward, the remaining elements that did not undergo changes in oxidation numbers should be balanced.
- The last step consists of balancing both H and O atoms present on both sides of the equations by adding H_2O, H_3O^+, H^+ or OH^-, depending on the acidity of the reaction media.

This balancing process could be summarized as: oxidation numbers-cations-anions-hydrogens-oxygens[9].

5.2. Half-reaction method

Though electrochemical processes occur simultaneously, the overall reactions are often split into separate oxidation and reduction half-reactions to facilitate the balancing process. In redox chemistry, the number of electrons accepted always equals the number of electrons donated. Thus, the sum of both oxidation and reduction two half-reactions must always cancel the total number of electrons in the overall reaction. The following steps should be followed to balance redox reactions using the half-reaction method.

- The first step consists of eliminating the spectator ions that do not really participate in the reactions, such as Na^+, SO_4^{2-} and ClO_4^-, and focus on the important species participating in the process.
- Next, the overall reaction should be divided into two half-reactions, each of them should be balanced by maintaining the same number of atoms of each element on both sides of the equation. This is often achieved by adding water ions (H^+, H_3O^+, OH^-). Depending on the media. Specifically, species like (H^+, H_3O^+, H_2O) should be used in acidic media and (H_2O, OH^-) in alkaline media.
- Afterward, electrons could be added to balance the charge. To equalize the number of electrons, the half-reactions should be multiplied by appropriate coefficients.
- The last step consists of summing the two half-reactions and eliminating the number of electrons and other spectator species.

This balancing process could be summarized as: half reactions-overall reaction-uninvolved ions[9].

Summary

During redox reactions, oxidation and reduction occur, where electrons lost by one species during the oxidation are gained by other species during the reduction process, and the sum of both reduction and oxidation processes yields an overall reaction with zero net electrons. It has to be kept in mind that the electrons exchanged during these processes are those occupying the outer shell, weakly attached to the nucleus by electrostatic forces. During redox processes, the oxidation numbers of the involved species undergo changes, by either increasing or decreasing. The oxidized atoms which lost electrons during the process (oxidants or oxidizing agents) will show an increase in their oxidation states. By contrast, the reduced atoms which gained electrons (reductants or reducing agents) will depict a decrease in their oxidation numbers. Although both the oxidation and reduction half-reactions could be expressed independently, electrochemical processes always occur together where electrons lost by one substance are gained by another. Therefore, during redox processes, oxidizing agents become reducing agents and vice versa. The assignment of oxidation numbers to elements in substances or compounds should always begin with the elements with stable known oxidation numbers, summarized by the above nine rules. The knowledge of the oxidation states is one way to balance redox reactions. However, the half-reaction method could also be utilized to achieve this goal.

References

1. Schüring, J., Schulz, H. D., Fischer, W. R., Böttcher, J., Duijnisveld, W. H. (1999). Redox: Fundamentals, Processes and Applications, Springer-Verlag, Heidelberg.
2. Masterton, W. L.; Hurley, C. N. (2008), Chemistry: Principles and Reactions, chapter 17, Cengage Learning.
3. Karen, P.; McArdle, P.; Takats, J. (2016), Comprehensive Definition of Oxidation State (IUPAC Recommendations 2016), Pure and Applied Chemistry. 88 (10).
4. Loock, H. P. (2011), Expanded Definition of the Oxidation State, Journal of Chemical Education. 88 (3): 282-283.
5. Jensen, W. B. (2007), The Origin of the Oxidation-State Concept, Journal of Chemical Education, 84, 1418.
6. Jensen, W. B. (2011), Oxidation States versus Oxidation Numbers, Journal of Chemical Education. 88 (12): 1599-1600.
7. Whitten, K.W.; Galley K. D.; Davis R. E. (1992), General Chemistry, 4th ed., Saunders.
8. Petrucci R. H.; Harwood W. S.; Herring F. G. (2002), General Chemistry, 8th ed., Prentice-Hall, pp. 81.
9. Dickerson, R. E.; Gray, H. B.; Haight, G. P. (1979), Chemical principles. 3rd ed. The Benjamin/Cummings Publishing Company, Inc., Menlo Park, CA.

Section 3

Practical Questions and Problems with Solutions

A set of practical questions and problems with detailed solutions are provided to better understand the discussed concepts. The questions and problems range from simple to complex.

Q1. i) What is the goal for determining the oxidation number of elements in substances? ii) Determine the oxidation number of chlorine in ClO_3^-, sulfur in H_2SO_4, manganese in Mn_2O_3, and chromium in $K_2Cr_2O_7$.

Ans1. i) The oxidation number helps determining the number of electrons that could be lost or gained during a redox reaction. The identification of the oxidation numbers helps to correctly balance redox reactions.

ii) To determine the oxidation number of an element in a compound, first add the oxidation states of all elements in the compound: $Cl + 3(O) = -1$. Next, use the nine rules to identify the oxidation numbers of key elements, such as oxygen (-2). This gives: $Cl + 3(-2) = -1$. Finally, find the oxidation state of chlorine: $Cl = -1 + 6 = 5$.

The same method should be used for sulfur in H_2SO_4. First add the oxidation states of all elements in the compound: $2(H) + S + 4(O) = 0$. Next, use the nine rules to identify the oxidation states of the main elements. Here, the oxidation states of oxygen and hydrogen are -2 and +1, respectively. The replacement of these oxidation states gives: $2(+1) + S + 4(-2) = 0$. Hence, S has an oxidation state of +6 in H_2SO_4.

For the oxidation number of manganese in Mn_2O_3, first add the oxidation states of all elements in the compound to yield: $2(Mn) + 3(O) = 0$. Then, use the oxidation states of the known elements (oxygen is -2). Next, replace oxygen by -2 to yield: $2(Mn) + 3(-2) = 0$. The oxidation state of Mn in Mn_2O_3 is +3.

Similarly, for chromium in $K_2Cr_2O_7$, add the oxidation states of all elements in the compound $2(K) + 2(Cr) + 7(O) = 0$. One of the nine rules states that the oxidation states of oxygen and potassium are -2 and +1, respectively. Next, replace these values in the formula to yield: $2(+1) + 2(Cr) + 7(-2) = 0$. The oxidation number of Cr in $K_2Cr_2O_7$ is +6.

Q2. i) Briefly, define oxidizing and reducing agents. Could a compound play an oxidizing and reducing agent at the same time? Explain how. The reaction between sulfur (IV) oxide and nitrogen (IV) oxide leads to the formation of nitric oxide and sulfur trioxide according to the reaction:

$$SO_{2(g)} + NO_{2(g)} \rightarrow SO_{3(g)} + NO_{(g)}$$

ii) Is this a redox reaction? Explain why and identify the oxidizing and reducing agents.

Ans2. i) An oxidizing agent is able to remove electrons from other compounds and a reducing agent is capable of donating electrons to oxidizing agents. Yes, an oxidizing agent could become a reducing agent and vice versa, depending on the reaction conditions.

ii) Yes, the reaction is redox because the oxidation numbers of the elements changed during the reaction. The oxidation number of sulfur in $SO_{2(g)}$ is +4 and in $SO_{3(g)}$ is +6. This suggests an oxidation because the oxidation number increased. The oxidation number of nitrogen in $NO_{2(g)}$ is +4 and in $NO_{(g)}$ is +2. The decrease in the oxidation number indicates a reduction process.

In sum, the oxidizing agent is $NO_{2(g)}$ because it gained electrons and $SO_{2(g)}$ is the reducing agent because it lost electrons.

Q3. Determine the oxidants and reductants in the following reaction: $AlCl_3 + 3K \rightarrow Al + 3KCl$

Ans3: This overall reaction is composed of two redox reactions (oxidation and reduction).

Oxidation: $3K^0 \rightarrow 3K^+ + 3e^-$

Reduction: $Al^{3+} + 3e^- \rightarrow Al^0$

Thus, $AlCl_3$ acts as the oxidant because it gained electrons and K is the reductant because it lost electrons.

Q4. i) Identify the oxidation numbers of Fe in $Fe_{0.80}O$ and $[Fe(H_2O)_5(NO)]SO_4$.

ii) Calculate the oxidation number of Cu in the reaction: $CuSO_4 + 2NaOH \rightarrow Cu(OH)_2 + Na_2SO_4$. Is this a redox reaction?

Ans4. The oxidation number of oxygen (O) is often -2. Thus, $0.80 \times Fe - 2 = 0$. The oxidation number of Fe in $Fe_{0.80}O$ is 2.5.

The oxidation number of H_2O and NO are zero and the oxidation number of SO_4 is -2. Thus, the oxidation number of Fe in $[Fe(H_2O)_5(NO)]SO_4$ is +2.

ii) The oxidation number of Cu is +2 on both sides of the reaction. Since the oxidation state did not change, the reaction cannot be classified as redox but rather a chemical process.

Q5. Consider the overall redox reaction: $PbS + H_2O_2 \rightarrow PbSO_4 + H_2O$

Identify the oxidation and reduction half-reactions. Which species are the oxidants and reductants?

Ans5. The identification of the two half-reactions starts by writing down the redox couples. The two redox couples involved in this reaction are: $PbS/PbSO_4$ and H_2O_2/H_2O.

In PbS /PbSO$_4$, the oxidation state of S increased from -2 to +6, thus this couple should be the oxidation half-reaction. The balancing of the reaction is performed by adding H$_2$O, H$^+$ or OH$^-$ to the reaction.

PbS + 4H$_2$O → PbSO$_4$ + 4e^- + 4H$^+$

The second half-reaction involves the redox couple H$_2$O$_2$ /H$_2$O, where the oxidation number of oxygen decreased from -1 to -2. Thus, it is the reduction half-reaction. Similarly, the balancing of the reaction is achieved by adding H$_2$O, H$^+$ or OH$^-$ to the reaction.

(H$_2$O$_2$ + 2H$^+$ + 2e^- → 2H$_2$O) × 2

Note that the reduction half-reaction is multiplied by a factor of 2 to eliminate the number of electrons in the overall reaction.

The oxidant is H$_2$O$_2$ because it gained electrons and PbS is the reductant because it donated electrons to H$_2$O$_2$.

Q6. Determine the oxidation number of P in HPO$_4^{2-}$.

Ans6. The nine rules indicate that the oxidation number of O is -2 and that of H is +1. Thus, +1 + P + 4 (-2) = -2, which gives an oxidation state of +5 for P in HPO$_4^{2-}$.

Q7. i) In your view, what would happen if F$_{2(g)}$ is left to react with Cl$^-_{(aq)}$? Explain the phenomenon using redox reactions. ii) What would happen when Cl$_{2(g)}$ is bubbled in KI$_{(aq)}$ solution, and Br$_{2(g)}$ bubbled in KCl$_{(aq)}$ solution?

Ans7. i) First of all, the redox potentials of both couples F$_2$/F$^-$ and Cl$^-$/Cl$_2$ should be checked to compare their reactivities. The standard reduction potential of F$_2$/F$^-$ = 2.87 V vs. NHE and that of Cl$_2$/Cl$^-$ = 1.35 V vs. NHE. Thus, F$_2$/F$^-$ is more reactive and reduces Cl$^-$ into Cl$_2$.

The two half-reactions can be summarized as follows:

Reduction: F$_{2(g)}$ + 2e^- → 2F$^-_{(aq)}$

Oxidation: 2Cl$^-_{(aq)}$ → Cl$_{2(g)}$ + 2e^-

Overall reaction: F$_{2(g)}$ + 2Cl$^-_{(aq)}$ → 2F$^-_{(aq)}$ + Cl$_{2(g)}$

In sum, this reaction should produce chlorine gas.

ii) When Cl$_{2(g)}$ is bubbled in KI$_{(aq)}$ solution, a reaction will occur if one of the species is more reactive than the other. The standard reduction potential of Cl$_2$/Cl$^-$ = 1.35 V and that of I$^-$/I$_2$ = 0.53 V. Thus, chlorine should be the reductant since its potential is higher, and the two redox reactions can be summarized as.

Reduction: Cl$_{2(g)}$ + 2e^- → 2Cl$^-_{(aq)}$

Oxidation: $2I^-_{(aq)} \rightarrow I_{2(l)} + 2e^-$

Overall reaction: $Cl_{2(g)} + I^-_{(aq)} \rightarrow I_{2(aq)} + Cl^-_{(aq)}$

K^+ is a spectator ion because it is not involved in the reaction, thus can be added to both sides of the reaction to yield: $Cl_{2(g)} + KI_{(aq)} \rightarrow I_{2(aq)} + KCl_{(aq)}$

During this process, a change in color from purple into yellow will be observed.

Similarly, the reduction potential of the redox couple $Br_2/Br^- = 1.066$ V is lower than that of $Cl_2/Cl^- = 1.35$ V. Therefore, Br_2 is less reactive to react with Cl^- to induce a redox reaction. Consequently, no change in color of the solution will be observed.

Q8. What would happen if an $Mg_{(S)}$ rod is immersed in $CuSO_4$ solution? Explain the phenomenon using redox reactions.

Ans8. To figure out what will happen, the redox potentials of the two redox couples should be compared. The standard reduction potential of $Mg^{2+}/Mg = -2.372$ V vs. NHE and that of $Cu^{2+}/Cu = +0.337$ V vs. NHE. Because of its low potential, Mg will dissolve in the solution to form Mg^{2+} and generate electrons, which will be used to reduce Cu^{2+} into Cu that will deposit on the electrode. During this process, the initial blue color of $CuSO_4$ will vanish over time and brown solid will deposit on the Mg rod. The redox half-reactions and the overall reaction can be summarized as follows:

Oxidation: $Mg_{(s)} \rightarrow Mg^{2+} + 2e^-$

Reduction: $Cu^{2+} + 2e^- \rightarrow Cu$

Overall reaction: $Mg_{(s)} + Cu^{2+}_{(aq)} \rightarrow Cu_{(s)} + Mg^{2+}_{(aq)}$

SO_4^{2-} is a spectator ion that does not participate in the reaction and could be added to the final reaction to yield: $Mg_{(s)} + CuSO_{4(aq)} \rightarrow Cu_{(s)} + MgSO_{4(aq)}$

Q9. Consider the reaction between chlorine gas and bromine ions:

$Cl_{2(g)} + 2 Br^-_{(aq)} \rightarrow Br_{2(aq)} + 2Cl^-_{(aq)}$

i) Propose a set up to perform this reaction. ii) This reaction involves redox processes. Which species are the oxidizing and reducing agents and why?

Ans9. i) An aqueous solution of Br^- salt (e.g., KBr) could be put in a container, which then will be bubbled with chlorine gas ($Cl_{2(g)}$). The reactivity will depend on the redox potentials of the species. Because the reduction potential of ($Cl_2/Cl^- = 1.35$ V vs. NHE) is higher than that of ($Br_2/Br^- = 1.066$ V vs. NHE), Cl_2 will oxidize Br^-. The overall reaction could be split into two half-reactions.

Oxidation: $2\,Br^-_{(aq)} \rightarrow Br_{2(aq)} + 2e^-$

Reduction: $Cl_{2(g)} + 2e^- \rightarrow 2Cl^-_{(aq)}$

Thus, $Cl_{2(g)}$ is the oxidant that will collect (or gain) electrons and $Br^-_{(aq)}$ is the reductant that will donate (or give) electrons to $Cl_{2(g)}$. During this process, the oxidation number of Br changes from -1 in Br⁻ to 0 in Br_2, confirming that Br⁻ donates electrons to chlorine and is the reductant (or reducing agent).

Q10. Select the correct answers from each series of multiple propositions and explain why.

- The oxidation state reveals: i) loss of electrons from a given element, ii) gain of electrons by an element, iii) no electron transfer involved, and/or iv) loss or gain of electrons by an element.
- The oxidation state of S in S_8 is: i) 0, ii) -2, iii) 10, and/or iv) -4.
- The oxidation number of H in bond state is: i) +1, ii) (+1 or -1), iii) -5, and/or iv) -1.
- The oxidation number of oxygen in OF_2 is: i) -2, ii) 0, iii) +2, and/or iv) -1.
- The oxidation state of O in H_2O_2 is: i) -1, ii) -3, iii) +1, and/or iv) -2.

Ans10.

- The oxidation state of an element in a compound expresses a loss or gain of electrons. During redox processes, changes in the oxidation states occur. Thus, the correct answer is iv).
- One of the rules states that the oxidation numbers of uncombined elements and neutral substances are always zero. Thus, S in S_8 has an oxidation number of 0, and the correct answer is i).
- The oxidation number of H is often +1, except in metal hydrides (e.g., NaH), where it can have an oxidation state of -1. Therefore, the correct answer is ii).
- Oxygen often has an oxidation number of -2 and fluorine -1. Because fluorine F is more electronegative than oxygen O, it will keep its oxidation state and that of O has to be recalculated. Since OF_2 is neutral, the oxidation state of O in OF_2 is +2. The correct answer is iii).
- Oxygen often has an oxidation state of -2. In this particular case, the oxidation state of O is -1 since the oxidation state of hydrogen is +1. The correct answer is i).

Q11. Pick up the correct answers from the given propositions.

- The oxidation states of chlorine in Cl_2, NaCl and NaOCl are: i) (0, +1, -1), ii) (0, -1, +1), and/or (+1, 0, -1).
- The oxidation state of P in PCl_5 is: i) -5, ii) 0, and/or iii) +5.

- The oxidation state of O in BaO_2 is: i) -2, ii) 0, and/or iii) -1.
- The oxidation numbers of O in O_2^+ and O_2^- are: i) (+1/2 and -1/2), ii) (0 and -1), and/or iii) (-2 and -2).
- The oxidation state of N in HN_3 is: i) 0, ii) -1/3, and/or iii) -3.

Ans11.

- Cl_2 is a neutral species, thus the oxidation state of Cl in Cl_2 is 0. The first group of the periodic table includes Na with an oxidation state of +1 and since NaCl is neutral, Cl in NaCl will have an oxidation state of -1. Oxygen often has an oxidation state of -2 and Na +1. Therefore, the oxidation state of Cl in NaOCl is +1. The correct answer is ii).
- Cl has an oxidation state of -1 and since PCl_5 is neutral, the oxidation number of P in PCl_5 is +5. The correct answer is iii).
- The oxidation state of Ba is +2 and since BaO_2 is neutral, the oxidation number of O in BaO_2 is -1. The correct answer is iii).
- In O_2^+, two oxygen atoms share +1 charge and in O_2^-, two oxygens share -1 charge. Hence, the oxidation states of O in O_2^+ and O_2^- are +1/2 and -1/2, respectively. The correct answer is i).
- In HN_3, three N atoms share -1 charge, which gives an oxidation state of -1/3 for each N. The correct answer is ii).

Q12. Pick up the correct answers from the given propositions.

- For transition metals: i) free elements have oxidation states of 0, ii) the oxidation state equals to the charge of ions, and/or iii) oxidation state of neutral compounds is 0.
- Explain the reason of why Fe cannot form a +8 oxidation state though Ru and Os could.

Ans12.

- All the statements are correct for transition metals. Free elements (e.g., Fe, Ni) have oxidation states of 0. Metal ions (e.g., Fe^{2+}, Au^{3+}) have oxidation states of 2+ and 3+, respectively. The oxidation state of neutral compounds (e.g., $FeCl_3$, $CuCl_2$) is 0.
- Since Fe, Ru and Os occupy the same period of the periodic table, they all have the same number of electrons in the outer shell. Fe has 26 electrons with the electronic configuration of $[Ar]3d^64s^2$. Thus, if Fe used all its electrons, it can form oxidation states up to +8. However, due to its smaller size compared to Ru and Os, it can not accommodate higher positive charges.

Q13. Classify the following species by increasing order of oxidation state: i) N in (NO, N_2, HNO_2, NH_3, and HNO_3), ii) Cl in (HClO, Cl_2, ClO_2, and HCl), and iii) S in ($S_2O_8^{2-}$, H_2S, SO_2, S_8, and SO_4^{2-}).

Ans13. i) Before establishing an order, the oxidation state of N in each species should be determined and compared to the others. According to the main rules, the oxidation number of H is +1 and that of O is -2. Because all the species are neutral, the oxidation states of N in (NO, N_2, HNO_2, NH_3 and HNO_3) are (+2, 0, +3, -3, and +5), respectively. Therefore, the following order could be established: $HNO_3 > HNO_2 > NO > N_2 > NH_3$

ii) The same procedure should be applied to the second case by determining the oxidation states of Cl in all the species. The oxidation state of O is often -2 and that of H is +1. Thus, the oxidation states of Cl in (HClO, Cl_2, ClO_2, and HCl) are respectively (+1, 0, +4, and -1). In sum, the following order could be established: $ClO_2 > HClO > Cl_2 > HCl$.

iii) In the third series, the oxidation state of O is often -2 and that of H is +1. As a result, the oxidation states of S in ($S_2O_8^{2-}$, H_2S, SO_2, S_8, and SO_4^{2-}) are respectively (+7, -2, +4, 0, and +6). The following order could be established: $S_2O_8^{2-} > SO_4^{2-} > SO_2 > S_8 > H_2S$.

Q14. i) Briefly, define the oxidation state of an element in a compound. ii) What is the difference between oxidation and reduction? iii) Why is it important to balance chemical reactions? vi) Define oxidizing and reducing agents.

Ans14. i) The oxidation state (or number) expresses the degree of oxidation or loss of electrons by an element in a substance. ii) Oxidation expresses a loss of electrons and reduction a gain of electrons during redox processes. iii) The balancing of reactions in terms of mass and charge allows determining the stoichiometry of each species involved in the reaction and enable carrying correct calculations because both the mass and charge are always preserved in chemical processes (Lavoisier's law). iv) An oxidizing agent is able to remove electrons from a reducing agent. Therefore, reducing agents donate (or lose) electrons and oxidizing agents receive (or gain) electrons.

Q15. $BaCl_2$ reacts with H_2SO_4 to yield $BaSO_4$ and HCl. Determine the oxidation state of Ba in $BaCl_2$ and S in H_2SO_4. Write down a balanced reaction.

Ans15. The oxidation state of Cl is often -1 and those of O and H are respectively -2 and +1. In $BaCl_2$, (Ba) + 2 (-1) = 0. The oxidation state of Ba in $BaCl_2$ is +2.

Similarly, in H_2SO_4, 2 (+1) + (S) + 4 (-2) = 0. Consequently, the oxidation state of S in H_2SO_4 is +6.

The balanced reaction could be written as:

$BaCl + H_2SO_4 \rightarrow BaSO_4 + 2HCl$

Note that HCl is multiplied by a factor of 2 to balance the mass on both sides of the reaction. The charge on both sides is zero, meaning that the reaction is charge balanced.

Q16. Identify the oxidation state of zinc (Zn) in $ZnCO_3$ and cobalt (Co) in $CoBr_2$.

Ans16. The nine rules indicate that the oxidation state of O is -2 and that of C is + 4. Thus, Zn + 4 + 3(-2) = 0. The oxidation state of Zn in $ZnCO_3$ is +2.

In $CoBr_2$, the oxidation state of Br is -1. Hence, Co + 2(-1) = 0. The oxidation state of Co in $CoBr_2$ is +2.

Q17. Determine the oxidation states of the underlined elements in each compound: Na$\underline{I}$O$_3$, Al$_2$($\underline{S}$O$_4$)$_3$, Na$_2$$\underline{O}$$_2$, and Ca$\underline{H}$$_2$.

Ans15. In Na$\underline{I}$O$_3$, the oxidation state of Na is often +1 and that of O is -2. Hence, +1 + I + 3(-2) = 0. The oxidation state of I in Na$\underline{I}$O$_3$ is +5.

In Al$_2$($\underline{S}$O$_4$)$_3$, the oxidation state of Al is often +3 and that of O is -2. Hence, 2(+3) +3S + 12(-2) = 0. The oxidation state of S in Al$_2$($\underline{S}$O$_4$)$_3$ is +6.

In Na$_2$$\underline{O}$$_2$, the oxidation state of Na is +1. Therefore, 2(+1) + 2O = 0. The oxidation state of O in Na$_2$$\underline{O}$$_2$ is -1.

In Ca$\underline{H}$$_2$, the oxidation state of Ca is +2. Hence, +2 + 2H = 0. The oxidation state of H in Ca$\underline{H}$$_2$ is -1.

Q18. Iron (Fe) in presence of oxygen and moisture leads to corrosion of the metal according to the reaction.

$Fe + O_2 + H_2O \rightarrow Fe^{2+} + OH^-$

Write down the oxidation and reduction half-reactions and the overall balanced reaction. Which products are formed during the corrosion process?

Ans18. The first step is to identify the two redox couples. Here, Fe/Fe^{2+} and O_2/OH^-. Since the reaction is linked to corrosion, there is no need to verify the standard reduction potentials to figure out which couple will oxidize or reduce because oxygen is the oxidant in this case.

Oxidation: $(Fe \rightarrow Fe^{2+} + 2e^-) \times 2$

Reduction: $O_2 + 4e^- + 2H_2O \rightarrow 4OH^-$

Note that the reduction half-reaction involves 4 electrons, and to eliminate the number of electrons in the overall reaction, the oxidation reaction is multiplied by a factor of 2.

The overall balanced reaction could be written as: $2Fe + O_2 + 2H_2O \rightarrow 2Fe^{2+} + 4OH^-$

The product resulting from this reaction is mostly iron hydroxide ($Fe(OH)_2$). Note that corrosion could also lead to other hydroxide and/or oxide forms depending on the conditions.

Q19. Determine whether the reaction between H_2 and F_2 is a redox process.

$H_2 + F_2 \rightarrow 2HF$

Ans19. To determine whether the reaction is redox, the oxidation number of H and F should be identified on both sides of the reaction and compared. In the reactants side, both H and F have oxidation states of 0. In the products side, their oxidations states changed to +1 and -1, respectively. As a result, it can be concluded that the reaction is redox because the oxidation states of the elements in the compounds are altered.

Q20. Determine the oxidation states of Cr in CrO_4^{2-} and $HCr_2O_7^-$.

Ans20. The oxidation number of O is often -2 and that of H is +1. The overall charge of CrO_4^{2-} is -2. Thus, $Cr + 4(-2) = -2$. The oxidation state of Cr in CrO_4^{2-} is +6.

In $HCr_2O_7^-$, the overall charge of the ion is -1. Hence, $+1 + 2(Cr) + 7(-2) = -1$. The oxidation state of Cr in $HCr_2O_7^-$ is +6 as well.

Q21. i) Is the following reaction a redox process: $2Ag + Cl_2 \rightarrow 2AgCl$?

ii) If so, why? iii) Determine the reductants and oxidants.

Ans21. i) To determine whether the reaction is redox, the oxidation numbers of Ag and Cl should change. In the reactants side, both Cl and O have oxidation states of 0. In the product side, their oxidations states changed to +1 and -1, respectively. Since the oxidation states are altered, the reaction is redox.

ii) Since the oxidation state of Ag increased, Ag is oxidized to Ag^+. By contrast, since the oxidation state of Cl_2 decreased, Cl_2 is reduced to Cl^-. The two half-reactions could be summarized as follows:

Oxidation: $(Ag \rightarrow Ag^+ + e^-) \times 2$

Reduction: $Cl_2 + 2e^- \rightarrow 2Cl^-$

To eliminate the electrons in the overall reaction, the oxidation reaction is multiplied by a factor of 2 to yield: $2Ag + Cl_2 \rightarrow 2AgCl$

The oxidant is the species that gained electrons (Cl_2) and the reductant is the species that donated electrons (Ag).

Q22. Determine whether the proposed reactions involve redox processes. If so, explain why and determine the oxidant and reductant species, as well as the oxidation and reduction reactions.

$2Na_{(s)} + Cl_{2(g)} \rightarrow 2NaCl_{(s)}$

$CH_{4(g)} + 2O_{2(g)} \rightarrow CO_{2(g)} + 2H_2O_{(g)}$

$Mg^{2+} + Cu \rightarrow Mg + Cu^{2+}$

Ans22. To determine whether the reactions involve redox processes, the oxidation numbers of the elements in each compound must be calculated.

In $2Na_{(s)} + Cl_{2(g)} \rightarrow 2NaCl_{(s)}$, both Na and Cl_2 have oxidation states of 0 in each side. In the product side, their oxidations states changed to +1 and -1, respectively. Since the oxidation states are altered, the reaction is redox. This overall reaction could be split into two-half oxidation and reduction reactions.

Oxidation: $2Na \rightarrow 2Na^+ + 2e^-$

Reduction: $Cl_2 + 2e^- \rightarrow 2Cl^-$

Na is the species that lost electrons or the reductant and Cl_2 is the species that gained the electrons or the oxidant.

In $CH_{4(g)} + 2O_{2(g)} \rightarrow CO_{2(g)} + 2H_2O_{(g)}$, the oxidation state of C changes from -4 to +4 and that of O from 0 to -2. Therefore, the reaction is also redox. This overall reaction could be split into two-half oxidation and reduction reactions. To mass balance the reaction, H_2O and H^+ are added to both sides, and to eliminate the number of electrons in the overall reaction, the second half-reaction is multiplied by a factor of 2.

Oxidation: $CH_{4(g)} + 2H_2O \rightarrow CO_{2(g)} + 8e^- + 8H^+$

Reduction: $(O_2 + 4H^+ + 4e^- \rightarrow 2H_2O_{(g)}) \times 2$

The species that lost electrons is $CH_{4(g)}$ or the reductant and the species that gained electrons is O_2 or the oxidant.

In $Mg^{2+} + Cu \rightarrow Mg + Cu^{2+}$, the oxidation state of Mg changed from +2 to 0 and that of Cu from 0 to 2+. Hence, the reaction is also redox, which could be split into two half-reactions.

Oxidation: $Mg \rightarrow Mg^{2+} + 2e^-$

Reduction: $Cu^{2+} + 2e^- \rightarrow Cu$

The species that lost electrons is Mg (reductant) and that gained electrons is Cu^{2+} (oxidant).

Q23. Balance the following equation using the oxidation number and/or half-reaction methods.

$$Cr_2O_{3(s)} + Al_{(s)} \rightarrow Cr_{(s)} + Al_2O_{3(s)}$$

Ans23. The first step is to find out the oxidation states of each element on both sides of the reaction.

$$Cr_2O_{3(s)} + Al_{(s)} \rightarrow Cr_{(s)} + Al_2O_{3(s)}$$
$$32\text{-}0032\text{-}$$

Next, identify the species that lost electrons (oxidation half-reaction) and that gained electrons (reduction half-reaction).

The oxidation state of Cr decreased from +3 in $Cr_2O_{3(s)}$ to 0 in $Cr_{(s)}$. Therefore, there is a gain of 3 electrons (reduction half-reaction).

The oxidation state of Al increased from 0 in $Al_{(s)}$ to +3 in $Al_2O_{3(s)}$. Hence, there is a loss of 3 electrons (oxidation half-reaction).

The two half-reactions could be summarized as follows:

Oxidation: $Cr_2O_{3(s)} \rightarrow 2Cr_{(s)} + 6e^-$

Reduction: $2Al_{(s)} + 6e^- \rightarrow Al_2O_{3(s)}$

Both reactions are mass and change unbalanced. The mass could be balanced by adding H_2O and H^+ on both sides.

Oxidation: $\quad\quad\quad Cr_2O_{3(s)} + 6H^+ \rightarrow 2Cr_{(s)} + 6e^- + 3H_2O$

Reduction: $\quad\quad\quad 2Al_{(s)} + 6e^- + 3H_2O \rightarrow Al_2O_{3(s)} + 6H^+$

The next step is to sum the two half-reactions, as well as eliminate the number of electrons and any other reoccurring species.

$Cr_2O_{3(s)} + 6H^+ + 2Al_{(s)} + 6e^- + 3H_2O \rightarrow 2Cr_{(s)} + 6e^- + 3H_2O + Al_2O_{3(s)} + 6H^+$

The balanced reaction could be written as: $Cr_2O_{3(s)} + 2Al(s) \rightarrow 2Cr_{(s)} + Al_2O_{3(s)}$

Note that simple reactions like this one could simply be balanced by trying few multiplication factors but the method cited above helps balancing most reactions (simple or complex). Therefore, students are highly advised to familiarize themselves with the method.

Q24. Balance the following reaction using the oxidation number and/or half reaction methods.

$$AgNO_3 + Cu \rightarrow Cu(NO_3)_2 + Ag$$

Ans24. The first step is to assign the oxidation state of each element on both sides of the reaction and identify the oxidation and reduction half-reactions.

$$AgNO_3 + Cu \rightarrow Cu(NO_3)_2 + Ag$$
$$1\ 5\ 2\text{-}\quad 0\quad\quad 2\ 5\ 2\text{-}\quad\quad 0$$

The oxidation state of Ag decreased from +1 in $AgNO_3$ to 0 in Ag. Thus, there is a gain of 1 electron (reduction half-reaction). The oxidation state of Cu increased from 0 in Cu to +2 in $Cu(NO_3)_2$. Hence, there is a loss of 2 electrons (oxidation half-reaction).

Oxidation: $\quad Cu^0 \rightarrow Cu(NO_3)_2 + 2e^-$

Reduction: $\quad (AgNO_3 + 1e^- \rightarrow Ag^0) \times 2$

The next step is to multiply the reduction reaction by a factor of 2 to yield the same number of electrons as the oxidation half-reaction.

The two half-reactions could now be added to eliminate the number of electrons and yield a balanced overall reaction.

$$2AgNO_3 + Cu \rightarrow Cu(NO_3)_2 + 2Ag$$

Q25. Balance the following reaction using the oxidation number and/or half-reaction methods.

$$Ag_2S + HNO_3 \rightarrow AgNO_3 + NO + S + H_2O$$

Ans25. The first step is to assign the oxidation state of each element on both sides of the reaction and identify the oxidation and reduction half-reactions.

$$Ag_2S + HNO_3 \rightarrow AgNO_3 + NO + S + H_2O$$
$$1\ 2\text{-}\quad 1\ 5\ 2\text{-}\quad\quad 1\ 5\ 2\text{-}\quad 2\ 2\text{-}\quad 0\quad 1\ 2\text{-}$$

The oxidation state of S increased from -2 in Ag_2S to 0 in S. Thus, there is a loss of 2 electrons (oxidation half-reaction). The oxidation state of N decreased from +5 in HNO_3 to +2 in NO. Therefore, there is a gain of 3 electrons (reduction reaction). The two half-reactions could be summarized as follows:

Oxidation: $(2NO_3^- + Ag_2S \rightarrow S + 2e^- + 2AgNO_3) \times 3$

Reduction: $(3H^+ + HNO_3 + 3e^- \rightarrow NO + 2H_2O) \times 2$

To eliminate the number of electrons in the overall reaction, the oxidation reaction is multiplied by a factor of 3, representing the number of electrons involved in the reduction reaction.

The last step is to sum both half-reactions to yield a balanced overall reaction.

$$3Ag_2S + 8HNO_3 \rightarrow 6AgNO_3 + 2NO + 3S + 4H_2O$$

Q26. Determine whether the reaction between HCl and NaOH is a redox process.

$$HCl + NaOH \rightarrow NaCl + H_2O$$

Ans26. To determine whether the reaction is redox, the oxidation numbers of Cl and O should be identified on both sides of the reaction and compared. In the reactants side, both Cl and O have oxidation states of -1 and -2, respectively. In the products side, their oxidations states remained the same. It can be concluded that the reaction is not redox because the oxidation states of the elements in the compounds remained unchanged.

Table of Content

	Discount offers	1
	Introduction	2
	Abstract	3
1.	Electrochemical or redox reactions	3
2.	Oxidation state or number	4
3.	Identification of oxidizing and reducing agent	4
4.	Assigning an oxidation state	5
5.	Balancing redox reactions	6
	5.1. Oxidation number method	6
	5.2. Half-reaction method	6

Summary	7
References	8
Practical Questions/Problems with Solutions	9
Table of content	23
About the author	25

www.ingramcontent.com/pod-product-compliance
Lightning Source LLC
Chambersburg PA
CBHW062238220526
45471CB00009B/3538